Hard Work

A Day with a Plumber

By Mark Thomas

Children's Press
A Division of Grolier Publishing
New York / London / Hong Kong / Sydney
Danbury, Connecticut

Photo Credits: Cover and photos by Maura Boruchow
Contributing Editor: Jeri Cipriano
Book Design: Christopher Logan

Visit Children's Press on the Internet at:
http://publishing.grolier.com

Library of Congress Cataloging-in-Publication Data

Thomas, Mark.
  A day with a plumber / by Mark Thomas
    p. cm. —(Hard work)
  Includes bibliographical references and index.
  ISBN 0-516-23138-3 (lib. bdg.)—ISBN 0-516-23063-8 (pbk.)
    1. Plumbing—Juvenile literature. 2. Plumbers—Juvenile literature. [1. Plumbers. 2.
 Occupations.] I. Title. II. Series.

TH6124 .T49 2000
696'.1'023—dc21
                                                                00-057032

Contents

My name is Paul.

I am a **plumber**.

I keep my tools in my van.

I use my tools to fix pipes.

This bathtub has a **clog**.

The clog blocks the water.

9

I use a special tool for
the clog.

The tool goes into
the **drain**.

The tool gets rid of the clog.

11

Next, I fix a **faucet**.

The faucet is on the kitchen **sink**.

13

I get under the sink.

I need to put in new pipes.

The pipes give water to
the faucet.

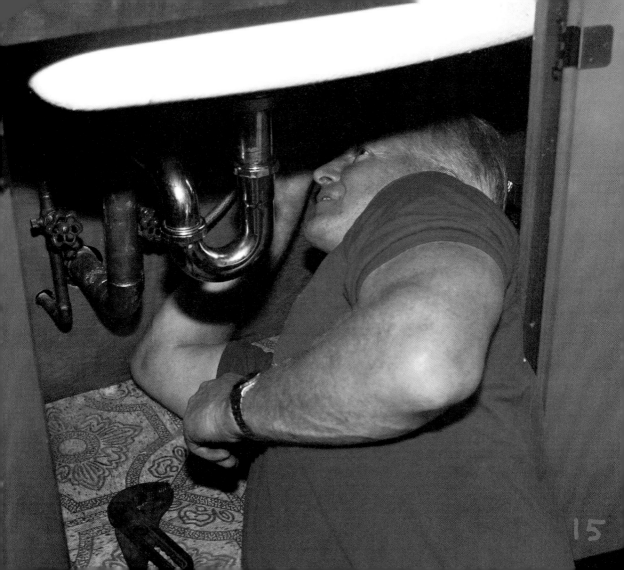

I use a **wrench** to keep the pipes together.

I turn the wrench to make a fit.

17

I turn on the water.

The water flows from
the faucet.

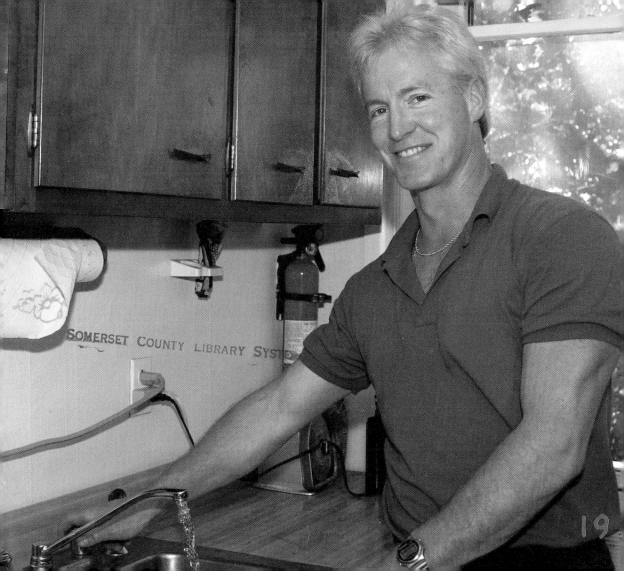

I put my tools back
in the van.

I like fixing things.

I like being a plumber.

New Words

clog (**klog**) something in a pipe that blocks the
   flow of water

drain (**drayn**) an opening for water to go
   into a pipe

faucet (**faw**-sit) the part on a sink that lets you turn
   on the water on and off

plumber (**plum**-er) a person who fixes pipes
   and plumbing

sink (**singk**) something that holds water and has a
   faucet that gives water

wrench (**rench**) a tool used to hold and
   turn pipes

To Find Out More

Book
I Can Be a Plumber
by Dee Lillegard
Children's Press

Web Site
Careers in Crafts
http:www.ncw.org.uk/careers/craftspeople/crafts.htm
Here you can read about different kinds of workers,
including plumbers.

Index

About the Author
Mark Thomas is a writer and educator who lives in Florida. He has built and repaired things in and around his home most of his life.

Reading Consultants
Kris Flynn, Coordinator, Small School District Literacy,
    The San Diego County Office of Education
Shelly Forys, Certified Reading Recovery Specialist,
    W.J. Zahnow Elementary School, Waterloo, IL
Peggy McNamara, Professor, Bank Street College of Education,
    Reading and Literacy Program